Minibeasts

Lynn Huggins-Cooper

A+

First published by Hodder Wayland
338 Euston Road, London NW1 3BH, United Kingdom
Hodder Wayland is an imprint of Hodder Children's
Books, a division of Hodder Headline Limited.

Design: Perry Tate Design, Language consultant: Andrew Burrell,
Science consultant: Dr. Carol Ballard

Published in the United States by Smart Apple Media
1980 Lookout Drive, North Mankato, MN 56003

Library of Congress Cataloging-in-Publication Data

Huggins-Cooper, Lynn.
Minibeasts / Lynn Huggins-Cooper. p. cm. — (Starters)
Summary: A simple introduction to some of the many different kinds of insects and
spiders that exist.
ISBN 1-58340-263-2 1. Insects—Juvenile literature. 2. Arachnida—Juvenile literature.
[1. Insects. 2. Spiders.] I. Title. II. Series.
QL467.2.H83 2003 595.7—dc21   2003041647

9 8 7 6 5 4 3 2 1

The publishers would like to thank the following for permission to reproduce
photographs in this book: Papilio; cover, contents page, 7, 9 (left), 17 / Corbis; 4, 18
(top), 22, 24 (second from top) / Premaphotos; 5 (bottom), 10 (top), 15 (bottom), 24
(fourth from top) / Oxford Scientific Films; 6, 8, 13 (bottom), 14, 16 (bottom), 19, 24
(bottom) / Heather Angel; 9 (right), 10 (bottom), 11, 18 (bottom), 21 (bottom), 24
(third from top) / NHPA; 12-13 (top) / Natural Visions; 16 (top) / Science Photo
Library; 20-21 (top), 23 / Getty; title page, 8, 24 (top)

Contents

Minibeasts everywhere!

Creeping, flying, swimming, LEAPING —wherever you are, minibeasts are nearby.

Some minibeasts are insects. Insects always have six legs. If a minibeast does not have six legs, it is not an insect.

Ladybugs have six legs, so they are insects.

Stag beetles
have six legs, so
they are insects.

Spiders have eight legs,
so they are NOT insects.

Bug hunt

Gardens all over the world are full of wildlife. Minibeasts are SMALL, but they can be fierce hunters! Many are predators.

Dragonflies catch flying insects. Their babies eat water insects, tadpoles, and even small fish!

Spiders eat flies and bugs. Sometimes they catch them in their sticky webs.

A praying mantis sits very still, waiting for a bug to come along—and then it pounces!

Copycats

Sometimes minibeasts look like something else. They may be camouflaged—the same color as their surroundings.

Leaf insects look exactly like leaves!

Camouflage helps keep minibeasts safe from predators—creatures that would eat them. It also helps them to hunt!

A flower mantis looks like a beautiful flower—until it pounces!

Stick insects look like dry old twigs.

Some baby minibeasts look just like their parents, only SMALLER!

Baby snails look the same as their parents.

Other minibeasts change as they grOW. They hatch from eggs and are called "larvae."

Caterpillars are baby butterflies and moths. They are larvae.

They will go through a complete change before they become adults. We call this "metamorphosis."

This moth was once a caterpillar.

Monster minibeasts

Most minibeasts are quite small. But some are **HUGE**!

A dragonfly that lived at the same time as the dinosaurs had wings as **BIG** as a seagull's!

Giant water bugs, found in America and Australia, can be as big as hamsters!

The Queen Alexandra's birdwing butterfly is larger than many birds.

Giant African land snails
can be as big as
your hand!

Amazing minibeasts

Minibeasts can do amazing things!

A puss moth caterpillar has a fierce "face" marking behind its head. It uses it to scare away predators. This caterpillar can spit stinging liquid, too!

Dragonflies have special eyes that help them see in all directions.

Stink bugs put their bottoms in the air and squirt smelly liquid when they are scared—yuck!

Master builders

Many minibeasts build homes for themselves.

Inside a hive or nest, bees make combs from wax. They store honey in the combs. They keep their babies in them, too!

A caddisfly larva makes a covering from pebbles, sticks, or even snail shells.

Termites build enormous mounds from sand, clay, and spit. Sometimes they can be as tall as a house!

Helpful minibeasts

Some minibeasts are very helpful. Ladybugs and lacewings eat pests that harm plants.

Lacewings are a gardener's friend!

Earthworms help gardeners, too. As they travel through the soil, they make it rich and airy. This helps plants grow.

Other minibeasts, such as bees, make things that people can use.

Bees make sweet, tasty honey.

Harmful minibeasts

Some minibeasts, such as black widow spiders, are harmful because they have a poisonous bite.

When mosquitoes bite, they can make people very ill with diseases such as malaria.

Other minibeasts are harmful because they spread disease or damage crops or even buildings!

Woodworms burrow into wood and damage it.

Huge swarms of locusts eat and destroy crops growing in fields.

Be a minibeast explorer!

Minibeasts are found all over the world—in gardens, parks, and backyards. With a magnifying glass, you can be a minibeast explorer!

Remember to be careful about touching minibeasts if there are poisonous bugs where you live! And always return minibeasts to their natural homes.

When you go outside, what will you find?

Glossary and index

Camouflage A way of disguising a creature to look like something else. **8-9**

Insects Creatures with six legs, such as ladybugs and butterflies. **4-5**

Larvae Baby insects, which often look like worms or maggots. **10**

Metamorphosis The change from a baby minibeast to an adult. **11**

Minibeasts Small creatures such as insects, worms, and snails. They are also called bugs. **4-5**

Poisonous Containing or using a substance that causes death or harm. **20, 23**

Predators Hunters that catch and eat other creatures. **6, 9, 14**

Swarms Large groups of flying insects. **21**